Collins

INTERNATIONAL

Maths
Foundation
Activity Book B

T0173334

Published by Collins
An imprint of HarperCollins*Publishers*
The News Building, 1 London Bridge Street,
London, SE1 9GF, UK

HarperCollins*Publishers*
Macken House, 39/40 Mayor Street Upper,
Dublin 1, D01 C9W8, Ireland

Browse the complete Collins catalogue at
www.collins.co.uk

© HarperCollins*Publishers* Limited 2021

10 9 8 7 6 5 4 3

ISBN 978-0-00-846878-1

British Library Cataloguing-in-Publication Data
A catalogue record for this publication is available from the British Library.

Author: Peter Clarke
Publisher: Elaine Higgleton
Product manager: Letitia Luff
Commissioning editor: Rachel Houghton
Edited by: Sally Hillyer
Editorial management: Oriel Square
Cover designer: Kevin Robbins
Cover illustrations: Jouve India Pvt Ltd.
Internal illustrations: Jouve India Pvt. Ltd.
Typesetter: Jouve India Pvt. Ltd.
Production controller: Lyndsey Rogers
Printed and bound in India by
Replika Press Pvt. Ltd.

Acknowledgements

With thanks to all the kindergarten staff and their schools around the world who have helped with the development of this course, by sharing insights and commenting on and testing sample materials:

Calcutta International School: Sharmila Majumdar, Mrs Pratima Nayar, Preeti Roychoudhury, Tinku Yadav, Lakshmi Khanna, Mousumi Guha, Radhika Dhanuka, Archana Tiwari, Urmita Das; Gateway College (Sri Lanka): Kousala Benedict; Hawar International School: Kareen Barakat, Shahla Mohammed, Jennah Hussain; Manthan International School: Shalini Reddy; Monterey Pre-Primary: Adina Oram; Prometheus School: Aneesha Sahni, Deepa Nanda; Pragyanam School: Monika Sachdev; Rosary Sisters High School: Samar Sabat, Sireen Freij, Hiba Mousa; Solitaire Global School: Devi Nimmagadda; United Charter Schools (UCS): Tabassum Murtaza; Vietnam Australia International School: Holly Simpson

The publishers gratefully acknowledge the permission granted to reproduce the copyright material in this book. Every effort has been made to trace copyright holders and to obtain their permission for the use of copyright material. The publishers will gladly receive any information enabling them to rectify any error or omission at the first opportunity.

Extracts from Collins Big Cat readers reprinted by permission of HarperCollins *Publishers* Ltd

All © HarperCollins*Publishers*

MIX
Paper | Supporting
responsible forestry
FSC™ C007454

This book is produced from independently certified FSC™ paper to ensure responsible forest management.

For more information visit:
www.harpercollins.co.uk/green

Count and match

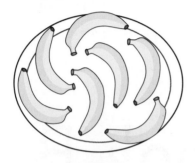

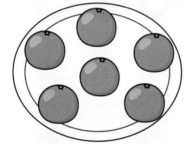

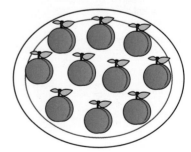

Draw lines to match plates with the same number of fruit.

Date:

Count and match

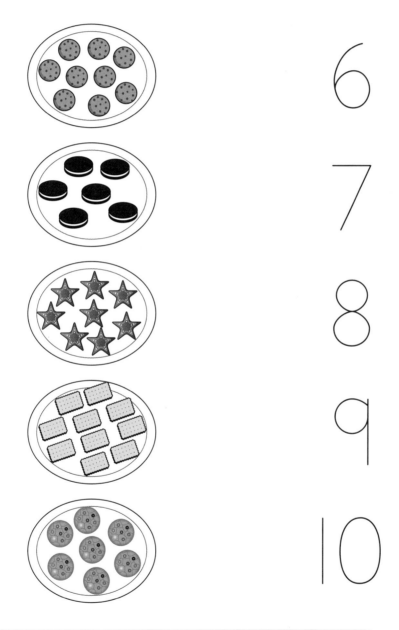

Count the biscuits on each plate. Draw a
line to the matching number.   Date:

Trace and write

6 6 6 6

7 7 7 7

8 8 8 8

9 9 9 9

10 10 10 10

Trace the numbers. At the end of each row,
write the number once more.   Date:

Trace and draw _ _ _ _ _ _

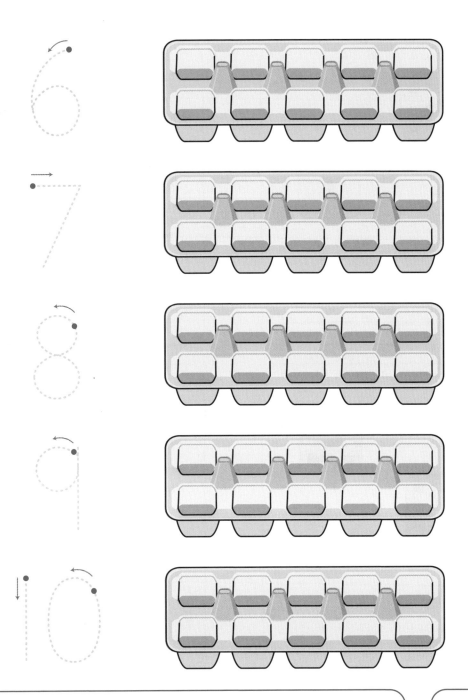

Trace each number. Draw the matching
number of eggs in the carton.

Date:

Add

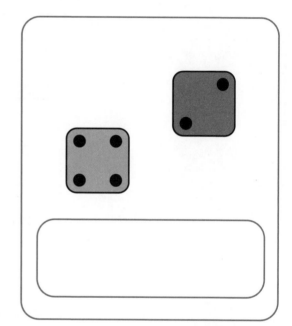

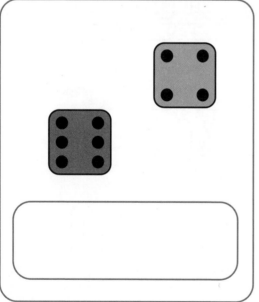

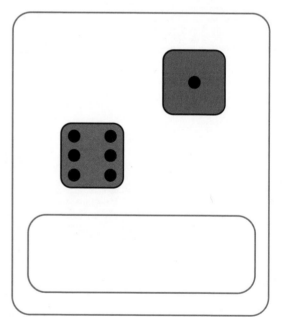

For each pair of dice, draw marks to show
the total number of dots.   Date:

Add

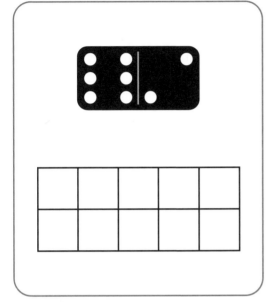

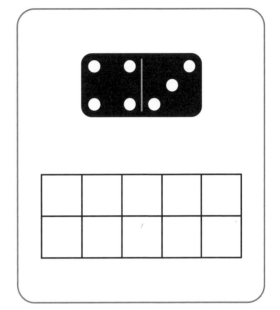

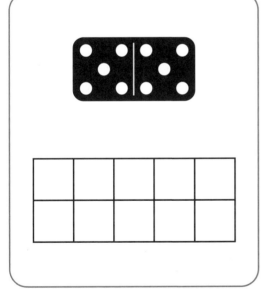

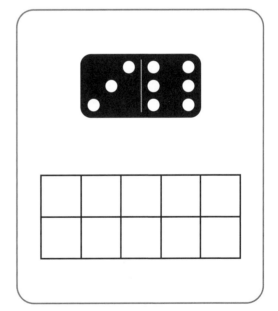

Draw dots in two colours on the ten-frame
to match the dots on the domino.

Date:

Add

1	2	3	4	5	6	7	8	9	10

Write the number of frogs. Write the number of fish.
Write the total.   Date:

Add

| 1 | 2 | 3 | 4 | 5 | 6 | 7 | 8 | 9 | 10 |

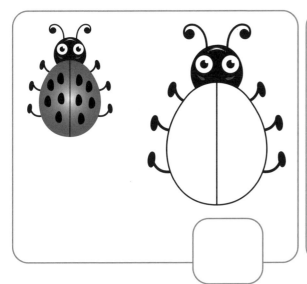

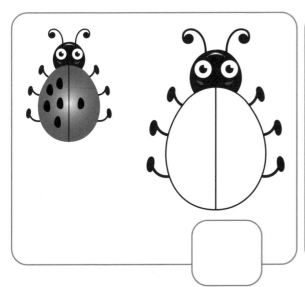

Count the spots on each wing. Write the numbers on the second ladybird. Write the total number of spots in the box.

Date:

I more

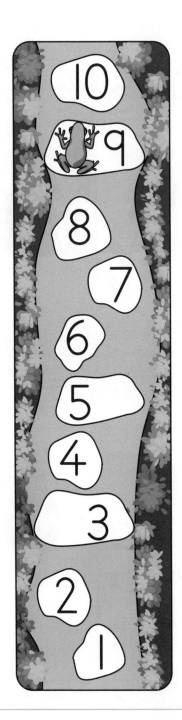

The frog jumps 1 more stone. Colour the
stone he lands on next.

Date:

2 more

| 1 | 2 | 3 | 4 | 5 | 6 | 7 | 8 | 9 | 10 |

For each cloud, count the raindrops. Draw
2 more. What is the total number of raindrops
for each cloud now?

Date:

Add on

| 1 | ②| 3 | 4 | 5 | 6 | 7 | 8 | 9 | 10 |

| 1 | 2 | 3 | ④| 5 | 6 | 7 | 8 | 9 | 10 |

| 1 | 2 | 3 | 4 | 5 | ⑥| 7 | 8 | 9 | 10 |

| 1 | 2 | 3 | 4 | ⑤| 6 | 7 | 8 | 9 | 10 |

Start with the circled number. Count on the number
of fingers. Colour the total on the number track.   Date:

Add more

Count the red fish. Circle that number. Count on
the number of blue fish. Colour the total.   Date:

Patterns

Continue each pattern.
Draw your own pattern.

Date:

Patterns

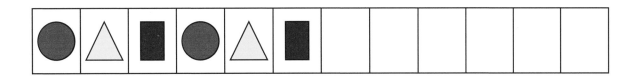

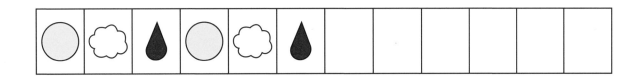

Continue each pattern.
Draw your own pattern.   Date:

Patterns

Continue each pattern.
Draw your own pattern.

Date:

Sort

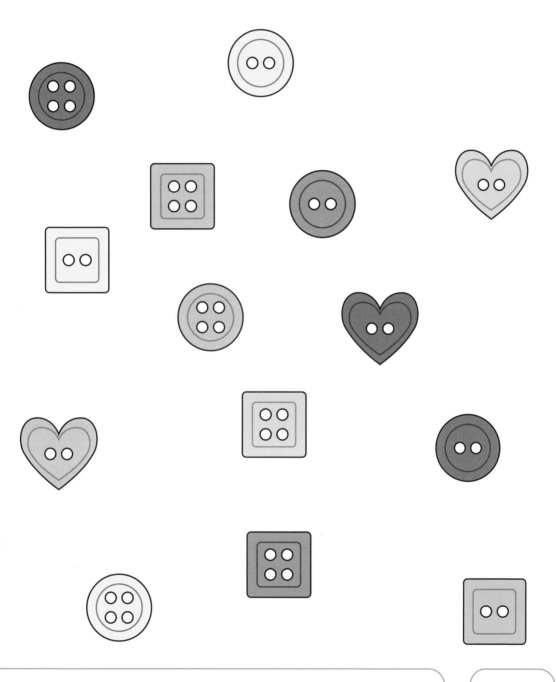

Choose a sorting rule (for example, 'buttons with
2 holes'). Tick the buttons that belong in that set.   Date:

Morning and night

Draw something you do in the morning.
Draw something you do at night.

Date:

Days of the week

Choose two days of the week. Ask an adult to write the days in the boxes. Draw something you like to do on each day.

Date:

Today, yesterday, tomorrow

Monday

Tuesday

Wednesday

Thursday

Friday

Saturday

Sunday

yesterday

today

tomorrow

Draw a line to match the correct day to
'today'. Repeat for 'yesterday' and 'tomorrow'.
Draw something you did/might do on each day.   Date:

Luke's exercise — diary

| Monday |

| Tuesday |

| Wednesday |

| Thursday |

Friday

Saturday

Sunday

Draw lines to match each day of the week with the exercise Luke did on that day.

Date:

Assessment record

_____ has achieved these Maths Foundation Phase Objectives:

Counting and understanding numbers

• Say and use the number names in order in familiar contexts such as number rhymes, songs, stories, counting games and activities, from 1 to 10.	1	2	3
• Say the number names in order, continuing the count on or back, from 1 to 10.	1	2	3
• Count objects from 1 to 10.	1	2	3
• Count in other contexts such as sounds or actions from 1 to 10.	1	2	3

Reading and writing numbers

• Recognise numbers from 1 to 10.	1	2	3
• Begin to record numbers, initially by making marks, progressing to writing numbers from 1 to 10.	1	2	3

Comparing and ordering numbers

• Use language such as more, less or fewer to compare two numbers or quantities from 1 to 10.	1	2	3

Understanding addition and subtraction

• In practical activities and discussions begin to use the vocabulary involved in addition: combining two sets and counting on.	1	2	3
• Find 1 more than a number from 1 to 10.	1	2	3

Patterns and sequences

• Talk about, recognise and make simple patterns using real-world objects or visual representations.	1	2	3

Time

• Begin to understand and use the vocabulary of time, including the days of the week, yesterday, today, tomorrow, morning and evening.	1	2	3
• Sequence familiar events.	1	2	3

Statistics

• Sort, represent and describe data using real-world objects or visual representations.	1	2	3

1: Partially achieved   2: Achieved   3: Exceeded

Signed by teacher:
Signed by parent: Date: